CONTENTS

All words marked in **bold** can be found in the glossary.

LOOKING AT COASTLINES

No two stretches of coast are exactly the same. There may be sheer cliffs, pebbly beaches, rock pools, muddy river **estuaries** and sand dunes, all within a small area.

The combination of land and sea provides a range of **habitats**. Each type of shore is home to a rich assortment of plants and animals.

Some coasts change along with the year's seasons. In **temperate** areas, a seashore that is crowded with windsurfers, yachts and fishing boats in the summer may be full of wading birds in the winter. Many coasts, especially in the **tropics**, are ravaged by fierce storms.

People come to the coast for fishing, trading, industry, vacations, water sports and to observe wildlife. It is no wonder that many major towns are on the world's coastlines.

SAFE HARBOR

Deep sheltered bays that are easy to reach from the open sea make good harbors. New York, San Francisco, and Sydney are all built on natural harbors.

▲ Hundreds of birds, animals, and plants live, feed, and breed on the remotest beaches. These pelicans scoop fish up from the sea into their special beak pouches. Pelican chicks put their heads right into their parent's beak pouches to pull food out.

PATTERNS ON THE COAST

The sea wears the coastline into many shapes. When waves crash on soft rock, it eventually falls away and then leaves bays, coves, and inlets. Harder rock is left as headlands, which are points of land that jut out into the sea.

Beaches are made of crushed rock thrown up on to the shore as sand or pebbles. Where the coast is flat, winds can push sand into hills, or dunes.

In Norway, Alaska, New Zealand, and Chile, the Ice Ages produced steep **fjords**. From the air, some river **deltas** look like a hand or fan, because of the deposited mud.

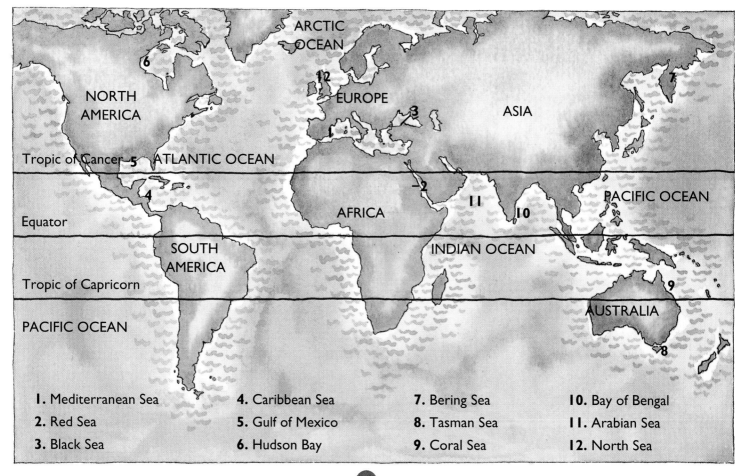

I. Mediterranean Sea	4. Caribbean Sea	7. Bering Sea	10. Bay of Bengal
2. Red Sea	5. Gulf of Mexico	8. Tasman Sea	11. Arabian Sea
3. Black Sea	6. Hudson Bay	9. Coral Sea	12. North Sea

CAVES AND CREEKS

Deep caves are formed by waves making a tunnel at a soft spot in a cliff.

Sea spray may explode through a blowhole in the roof of the cave.

Sooner or later, the roof of the cave may fall in, and a narrow creek will develop.

◄ Sometimes two caves join under a headland to form an arch. When the roof finally falls in, it leaves a stump of rock, or stack.

▼ Underwater **volcanoes** can throw up enormous piles of lava. These volcanic cliffs in Hawaii are some of the tallest in the world.

MOVING WATERS

The sea never stops moving. Every few seconds a wave breaks on the shore. Waves are put in motion by the wind, earthquakes, and the pull of the sun and moon.

Twice a day the **tide** comes in and goes out. At high tide, the sea surges onto the shore. It is drawn by the moon, which acts on the ocean waters like a magnet. At low tide, the sea swings back from the shore. When there is a full or new moon, the tides are higher.

Once in a while a vast tidal wave, or tsunami, the size of a four-story building, hits the shore in some tidal areas. Tsunamis can travel as far as 9,950 miles (16,000 km) a day and are caused by underwater earthquakes or volcanoes.

Strong winds around the Earth create **ocean currents**, which flow like rivers, carrying warm water from the tropics and cold water from the poles. A message in a bottle might travel from Mexico to Norway, carried by the circulating current.

SURVIVING THE TIDES

High-tide mark

1 Sea squirts, starfish, and periwinkles live near the high-tide mark.

Low-tide mark

2 Barnacles, limpets, crabs, and worms stay between the high- and low-tide marks. They are washed by the tide once or twice a day.

3 Cockles and razor shells prefer to live on the wetter lower shore.

▶ Some waves are so strong that they carry surfers right up to the beach before breaking. Surfers use the power at the edge of a wave. Local people even surfed off the coast of Hawaii before Christopher Columbus arrived in 1492.

▼ At low tide, the waves leave rippling patterns in the mud. Around the huge Atlantic and Pacific Oceans, tides go much further out and in than on the shores of the Mediterranean Sea. The larger the ocean is, the higher the tide is.

SEASHORE LIFE

The seashore is alive with all sorts of creatures. Most stay hidden from the waves, the sun and **predators**.

In the rock pools live shrimps, sea urchins, anemones, and starfish. On exposed rocks live barnacles and limpets. Shells protect their soft, wet bodies and help to keep in moisture.

On sandy beaches are the burrows of cockles, sea snails and all kinds of worms. Fat lugworms leave coils of sand like spaghetti on the surface.

The mud at river mouths teems with snails, fish and oysters that can survive in both fresh and salty water.

Most seashore animals leave their hiding places only at high tide to feed. Sea anemones use poisonous stings in their tentacles to catch food. Sea slugs eat the sea anemones and re-use these stings. Sea urchins have a sharp casing of spikes, while crabs use their strong pincers and claws for protection.

▼ Marine iguanas are the only lizards adapted to life by the sea. They swim with their tails and steer with their legs. When they come up for air, they blow salt out through their noses.

DID YOU KNOW?

● The American oyster lays up to one million eggs each time it **spawns** to increase the chance that some of its offspring will survive.

● Green turtles lay their eggs in a burrow on a sandy beach. When the eggs hatch, the baby turtles scuttle down by night to the safety of the dark sea. They know how to find their way by instinct.

● The grunion fish lays its eggs at low tide. Two weeks later, when the eggs are ready to hatch at sea, the high tide sweeps in and loosens them.

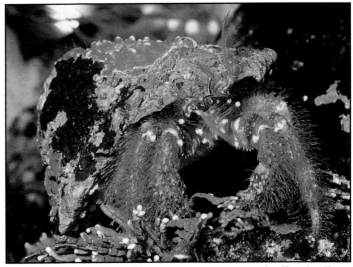

▲ The hermit crab uses an empty whelk shell as a mobile home. An anemone will often hitch a ride on its shell and help keep predators away.

▼ A hungry killer whale charges up the shore towards a group of plump seals. The seals have little defense against this surprise attack.

BIRDS ON THE SHORE

Many birds are seasonal visitors to coastal cliffs and shores. Gulls stay almost all year, but sandpipers and other wading birds breed inland and only spend winter on the coast. Puffins, gannets and frigate birds feed out at sea, then come ashore to breed in huge, busy colonies.

Many species **migrate** in winter to the warm coasts of the south. Terns, petrels, shearwaters and skuas breed in the north, then fly thousands of miles to Africa, Australia, the Americas, and Antarctica.

Some birds are expert fishers. Auks can move quickly by flapping their wings underwater. Cormorants catch fish with their feet. Gannets can dive into the sea from high in the sky to catch the fish underwater.

▲ Razorbills, fulmars and puffins take shelter in the cracks of a rocky cliff in the Shetland Islands, Scotland. Puffins usually nest in burrows, with up to one million pairs in one breeding colony. The parents can hold mouthfuls of fish to feed their young.

◀ An oystercatcher has a long, pointed beak. It pries shellfish from the surface of rocks and bores a hole in the shell. Its beak is even sharp enough to open up the tough shells of cockles and mussels.

▶ The male frigate bird puffs out his scarlet chest into a huge balloon to attract his mate. Even after mating, he will sometimes give a display.

▼ Gannets mate for life and have a kissing ceremony when they greet each other. Gannet chicks that hatch early in the breeding season are more likely to survive when they leave the nest than chicks that hatch later.

PLANT SURVIVORS

Plants on the coast need to be hardy to survive. They are blasted by salty winds, and the soil may be dry, wet, or just bare rock. Plants can be swamped by the tide, then roasted by the sun, or be buried by sand.

Some coastal plants have strong **sap**, which allows them to suck in salt water without losing their juice. On the salt marsh, matted cushions of thrift, sea lavender and glasswort spring up. On sand dunes, marram grass anchors the sand and shelters other plants, even little orchids.

Plants that tolerate wind and sea spray can even grow out of cliffs. On pebbly beaches with little soil, sea campion and sea peas find a home.

On the shoreline, seaweed comes in three colors: red in the shallows, brown in the middle of the beach, and green at the high-tide mark.

▼ Mangrove trees have strange roots suspended above the water which anchor the tree and help it to "breathe" in swampy coastal sea water.

FLOWER POWER

● The evening primrose grows on sand dunes. It opens its flowers in the evening when night-flying moths come out to feed.

● Seaweed is slimy and elastic to bend with the waves. Some species have bladders of air that hold branches up to the light under the surface of the water.

● Millions of tiny plants grow in the ocean, wherever they can receive light. Minute animals mix with the plants to form a rich "soup" called plankton. Some plankton will develop into larger animals or plants.

▶ Coconut palms grow well in the sandy soil on tropical beaches. If the nuts fall from a tree onto the shore, they may be picked up by the tide and then drift far out to sea. Sometimes, coconuts stay afloat for months, bobbing on the water, before they are washed up on a new and distant shore and take root.

COASTLINE PEOPLE

Native coastal people have a close relationship with the sea. The Inuit of Alaska can draw an accurate map of their shoreline by heart. Many coastal people can forecast weather from signs given by nature.

Seashore people respect marine life and take only what they need. On the Pacific coast of the U.S., locals protested the harm caused by commercial fishing. Now they manage the **fisheries** themselves.

More **Aboriginal** people live on Australia's coast than inland. They use traditional spears and fish traps to catch turtles and salmon, but sail in modern boats instead of canoes.

Tourism has changed life on the coast, especially in nations like Mexico and Jamaica. Some people work in tourism in the summer, and return to their traditional jobs in the winter. Others have totally abandoned their old way of life.

▲ A boy and his father bring home the morning catch of fish and lobster on the east coast of Malaysia. Many children who live on the coast in this part of the world learn to fish at an early age. Each culture has its own way of fishing. On coasts where there are few fish in the shallows, people must sail to deeper waters by boat. In places where fish are more abundant in the shallow coastal seas, people pull nets across a bay or use traps, baskets, spears, and rods.

◀ For some people, home is a boat, even in crowded coastal cities. Many of these cities are ringed by high mountains. There is not enough land to house the whole population, and so the poorest people live on boats on the sea. At least this bay in Hong Kong offers people and boats some shelter from **typhoons**.

PEOPLE FACTS

● It is still traditional in Portugal to paint eyes on boats. Fishermen once thought the eyes would help to locate and catch fish.

● On the Black Sea coast, people cover themselves in black mud. It is good for the skin and is said to help relieve aches.

● In Malaysia, some coastal people live in longhouses raised on stilts to keep them high above the **monsoon** floods. Up to 14 families may live in each house.

● All over the world, coastal people preserve fish by a smoking process. In the Scottish isles, herrings are slowly smoked over an oak and sawdust fire to make kippers.

SEASHORE RICHES

The world's shores are brimming with treasures. Two thirds of all fish caught are hatched in coastal waters. Commercial fishing brings in huge catches – over 80 million tons each year. But there is a danger of over-exploiting some species. To save them from extinction, herrings can no longer be caught in the North Sea, by law. There is also a ban on catching sardines off California. Many countries put a fishing limit around their shores to protect fish supplies from foreign fishing fleets.

The most profitable minerals from the sea are oil and natural gas. In the future, waves and tides may be used to generate electricity. There is already one tidal power station in France. But the sea's most precious gift is the unique life it supports.

▲ Two men at a floating market in Thailand trade fish and shellfish.

TREASURE TROVE

Many medicines are made from plants and creatures in the sea. Seaweed provides iodine to treat cuts and bruises.

Oil drilled from the seabed is refined into gasoline. Sea salt is used to flavor food, and sand is used to make cement.

Precious pearls form inside oyster shells. Oysters, other kinds of shellfish, and lobsters are bred on coastal fish farms.

▼ The tide sweeps up huge harvests of fish close to the shore. These fishermen rig up their complicated nets in a coastal inlet. As the tide rises, they will slowly lift their nets out of the water and scoop up a heavy catch.

DANGER ALERT

Coastlines are in danger of erosion, but people are the worst threat. Mangrove forests protect tropical coastlines and provide breeding grounds for marine life. In the Far East, these forests are cut down for housing, industry, and furniture.

Coral reefs struggle to survive under **silt**. Sea animals are put at risk from sewage, toxic waste from factories and farms, and oil spills. Migrating salmon are sensitive to warm water from power stations.

Even worse, air pollution is now causing the planet to heat up. This could melt the polar ice caps, and lead to flooding around the world. Low-lying coastal nations like The Netherlands could disappear.

▲ Trash on beaches can be dangerous. Every year, many seabirds and sea mammals are strangled or suffocated by plastic bags and cans. As this **scavenging** cow crosses the sand, it may damage itself on broken glass and jagged metal.

◀ Two Filipino girls sell dyed sea snake skins to be made into bags, belts, or shoes. Certain snakes native to coastlines may become extinct if they continue to be killed in large numbers.

▶ This seal is trapped in a drift net left by a fishing boat. The careless use of drift nets in commercial fishing kills thousands of seals, whales, and dolphins each year.

▼ A penguin stands on the shore, covered in oil from a slick. Oiled birds are unable to swim, catch fish for food, or stay warm.

SAVE THE COAST

Some countries have created marine nature reserves in their coastal waters. The Great Barrier Reef in Australia is the biggest. The government in Australia has also given the mangrove forests to the Aboriginal people for safekeeping.

In Europe, the Wadden Sea in The Netherlands and the Camargue in the South of France are havens for rare coastal species. However, there are still very few coastal reserves compared to parks inland. Conservation groups work hard to protect endangered species and to persuade companies to cut down on pollution. Much more still needs to be done.

▼ The future of these young albatross chicks lies in our hands. We must take good care of the coasts to help them survive.

WHAT YOU CAN DO

- Take trash home instead of leaving it on the beach.

- Leave nesting birds undisturbed.

- Refuse to buy shells and coral.

- Write to companies that cause pollution and ask them to stop.

- Join a conservation society.

LUTEY AND THE MERMAID

*For thousands of years, people have told stories about the world
around them. Often, these stories try to explain something
that people do not understand, such as how the world began.
This story, told by the people of Cornwall on Britain's coast,
is about a mermaid. Some people say stories of mermaids
developed from sightings of distant seals.*

Lutey was a fisherman who lived in Cornwall long ago. Every day, he would go out in his little boat to catch fish to sell at the market.

One day, as he was dragging his boat up on to the shore, he heard a strange cry. It was not the cry of any animal he could think of, nor was it the cry of a human being. The cry seemed to be coming from behind a large outcrop of rock. Leaving his boat, Lutey crept up to the rock and peered around.

There, in a large rock pool, was a mermaid. She was weeping and moaning and thrashing her silvery tail in the pool.

Lutey was very surprised. He had heard tell of mermaids, but he had never seen one. He stared with his mouth wide open.

"Oh, dear. Oh, dear," sobbed the mermaid. "Whatever shall I do?"

"W... What's the matter?" Lutey stammered.

The mermaid looked up and saw Lutey standing there.

"Oh, thank goodness you're here!" she cried, sweeping her hair back from her face. "I'm in the most terrible trouble. The tide has gone out and left me stranded in this rock pool. If I don't get back to the sea before supper time, my husband will eat the children."

"What a terrible husband you must have!" said Lutey, very shocked.

"He's a dear, really," the mermaid replied. "It's just that he does get so very hungry and the children are so small and tender that he really can't help himself."

Without further ado, Lutey picked up the mermaid and carried her to the edge of the sea. He laid her down carefully in the shallows.

The mermaid swished her tail and tossed her wet hair.

"You have been so kind, I hardly know how to thank you," she said.

"There's no need to thank me," said Lutey. "Just get off to that husband of yours, before he eats the children."

"I really must thank you," she insisted. "I will give you this comb, and if you ever want my help, just run the comb through the water three times."

With a flick of her silvery tail, the mermaid disappeared beneath the shimmering waves.

Lutey put the mermaid's comb in his pocket. He decided not to tell anyone about the mermaid. They would probably not have believed him anyway.

But, from that day on, people did notice something different about Lutey. Each time he took his boat out, he came back laden with fish, even on days when the other fishermen had caught nothing. And on stormy days, when no one else would venture out to sea, Lutey would sail out in his little boat, and come home with not a single tear in his sail. People began to say that he was charmed.

Of course, none of them knew that Lutey always carried a mermaid's comb in his pocket.

One day, exactly seven years after Lutey had met the mermaid, a great wind arose. Lutey was wandering along the shore, looking out to sea. Although it was quite safe for him to go out to sea in bad weather, he no longer needed to, because he had become rich from all his good hauls.

By the shore he met some fishermen, who were discussing whether or not to take their boats out in this wind.

"Good morning, Lutey," they said, as he neared them. "Do you think there's a bad storm brewing?"

"I'm sure I don't know," Lutey replied. "But if you are planning to go out today, I'll go with you."

The other fishermen were very pleased to hear this. Fishermen are very superstitious, and these men believed Lutey would bring them good luck.

So they all set out together in a big fishing boat. The strong wind blew them far from the shore very quickly, but then it calmed. The fishermen cast their nets and drew them in bursting with fish.

"This is wonderful!" they shouted. "You have brought us very good luck, Lutey."

With that, they set sail for home. But suddenly the storm arose again. The fierce wind battered the boat and the waves lashed against it.

"What shall we do?" they cried. "We will never reach home!"

"Don't despair," said Lutey, calmly. He made his way to the bow of the boat. Then he felt in his pocket and brought out the mermaid's comb. The other fishermen watched anxiously, but through the wind and rain, they could not clearly see what he was doing. They saw a flash of gold in Lutey's hand and then he leaned far forward over the bow. All of a sudden, Lutey seemed to drop forward into the water. All the fishermen rushed to the bow to try to rescue him, but Lutey had vanished. Far off on the port side, though, one of them saw a huge silver fishtail splashing through the foaming waves.

Gradually the weather calmed, but the fishermen could see no sign of their friend. Sadly, they sailed safely back to shore. And Lutey? Some people say that the mermaid came to fetch him when her husband died and now they live happily together at the bottom of the sea.

TRUE OR FALSE?

Which of these facts are true and which ones are false?
If you have read this book carefully, you will know the answers.

1. Pelicans are the heaviest birds in the world.

2. The tide goes in and out three times a day.

3. Some anemones live on the shells of hermit crabs.

4. Seaweed comes in three colors: yellow, blue and pink.

5. Some sand dunes in the south of France are as tall as twenty people.

6. Lugworms leave blobs of sand-like peas on the surface of the sand.

7. Iodine for treating cuts and bruises is made from sea squirts.

8. The Inuit people of Alaska know the shape of their coastline by heart.

9. Puffins nest in burrows, in colonies holding as many as one million pairs.

10. Marine iguanas steer with their tails and swim with their legs.

11. Fewer than 500 Mediterranean monk seals are left in the world.

12. In the Far East, coastal children learn to fish when they are twelve years old.

GLOSSARY

● **Aboriginal** people are the native inhabitants of a country. Often, the word is used to describe the native people of Australia.

● **Coral reefs** are bands of coral formed off a coastline. Coral is a limestone deposit formed by millions of tiny sea creatures called coral polyps. Some corals look like organ pipes, brains, or mushrooms. They come in shades of red, yellow, pink, brown, purple, and green.

● **Deltas** are where river mouths join the sea and break into small rivers with sand or silt deposits in between. They look like a net of land and water, and are often shaped like a triangle.

● **Erosion** is when land is gradually worn away by the sea, the wind, or human activities such as farming. Trees and plants help to protect soil from erosion.

● **Estuaries** are where rivers open out into the sea. Fresh and salt water mix there. Some estuaries are only narrow channels. Huge rivers, such as the Amazon, gape open for many miles at the coast.

● **Fisheries** are areas of sea where fish are known to be and where people harvest large catches of fish.

● **Fjords** are long, narrow inlets with steep cliffs. They were carved out by glaciers during the Ice Age.

● **Habitats** are environments to which certain plants and animals have adapted. The coast, cliffs, sand dunes, and rock pools are all habitats with unique conditions and different groups of plants and wildlife.

The Nile Delta

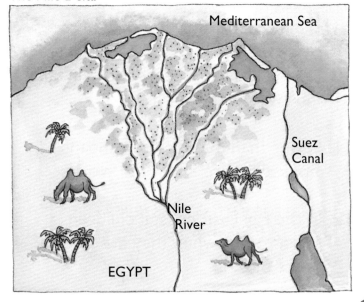

Mediterranean Sea

Suez Canal

Nile River

EGYPT

● **Migrate** is what some birds do when, every year, they fly a long way from their breeding place to winter feeding grounds. Sometimes they fly from one continent to another.

● **Monsoon** is the rainy season in Asia. It has strong winds and torrential rain, and sometimes brings flooding after the dry season.

● **Ocean currents** are strong movements of sea water. Each current flows along its own route. Currents usually move in huge circles between the continents.

● **Predators** are animals that hunt, kill, and feed off other animals.

● **Sap** is the juice inside a plant. It carries food and water to the leaves, the same way blood circulates food in the bodies of animals.

● **A scavenger** is an animal that feeds on the remains of food that has been left by other animals.

● **Silt** is a layer of fine mud dumped by moving water. When the sea bottom is scraped to harvest shellfish, or to clear a harbor, silt is stirred up.

● To **spawn** is to deposit a mass of eggs. Fish and other small sea creatures spawn.

● **Temperate** is the area of the Earth between the cold poles and the hot tropics. The climate is fairly warm, but varies with the seasons.

● The **tide** is the daily and monthly movement of the sea up and down the beach.

● The **tropics** are the part of the earth near the equator. The area is hot and often wet.

● **Typhoons** are violent storms with strong winds like a hurricane. They blow in circles and can destroy houses and knock down trees.

● **Volcanoes** are formed when melted rock from the center of the Earth shoots into the air, solidifies and forms cone-shaped hills.

INDEX

RESOURCES

At the Seashore, by Pamela Hickman, 1997. This book is filled with ideas of things to do and discover along the coastline.

Coral Reef, by Paul Fleisher, 1998. This book explores animals that make up the brightly colored underwater gardens called coral reefs.

Interfact Coral Reefs, by World Book, 1999. This CD-ROM and book work together to help students learn about coral reefs and the animals that live in them. The disk is full of interactive activities, puzzles, quizzes, games, and interesting facts. The book contains fascinating information highlighted with lots of full-color illustrations and photographs.

Seashore Life, by Barbara Taylor, 2000. This book, part of the Natural World Series, discusses all forms of marine life.